太 阳

太阳系的唯一恒星

THE SUN

The Star of Our Solar System

（英国）埃伦·劳伦斯／著　张　骁／译

江苏凤凰美术出版社

著作权合同登记图字：10-2022-144

图书在版编目（CIP）数据

太阳：太阳系的唯一恒星 / (英) 埃伦·劳伦斯著；

张骁译. –– 南京：江苏凤凰美术出版社, 2025. 1.

(环游太空). –– ISBN 978-7-5741-2027-3

Ⅰ. P182-49

中国国家版本馆CIP数据核字第2024ZW4995号

策　　　　划	朱　婧
责 任 编 辑	高　静　奚　鑫
责 任 校 对	王　璇
责任设计编辑	樊旭颖
责 任 监 印	生　嫄
英 文 朗 读	C.A.Scully
项 目 协 助	邵楚楚　乔一文雯

丛　书　名	环游太空
书　　　名	太阳：太阳系的唯一恒星
著　　　者	（英国）埃伦·劳伦斯
译　　　者	张　骁
出 版 发 行	江苏凤凰美术出版社（南京市湖南路 1 号 邮编：210009）
印　　　刷	南京新世纪联盟印务有限公司
开　　　本	710 mm×1000 mm　1/16
总 印 张	18
版　　　次	2025 年 1 月第 1 版
印　　　次	2025 年 1 月第 1 次印刷
标 准 书 号	ISBN 978-7-5741-2027-3
总 定 价	198.00 元（全 12 册）

版权所有　侵权必究

营销部电话：025-68155675　营销部地址：南京市湖南路 1 号
江苏凤凰美术出版社图书凡印装错误可向承印厂调换

目录  Contents

书中加粗的词语见词汇表解释。
Words shown in **bold** in the text are explained in the glossary.

遇见太阳
Meet the Sun

在距地球上亿千米外，一颗巨大的燃烧着的火球在黑暗的宇宙中闪耀着。

Hundreds of millions of kilometers from Earth, a giant, burning ball of light shines in the blackness of space.

在地球上，我们能看见它的光亮，也能感受到它的热度。

Here on Earth, we see its light and feel its heat.

如果没有它的光和热，我们的世界将变得黑暗且冰冷。

Without its light and heat, our world would be dark and freezing cold.

也将没有任何生命可以在我们的母星上生存。

No life on our home **planet** could survive.

这个巨大的发光球体就是太阳，正是它使地球有了孕育生命的可能!

The huge, shining ball is the Sun, and it makes life on Earth possible!

请不要直视太阳，因为这会对视力造成严重损害。科学家们用特殊仪器拍摄了太阳的照片，比如这本书中的插图，这样所有人就都可以安全地看到这些奇妙的照片了。

You should never look directly at the Sun because it will badly damage your eyes. Scientists take photos of the Sun, like the ones in this book, using special equipment. Then everyone can safely look at the amazing photos.

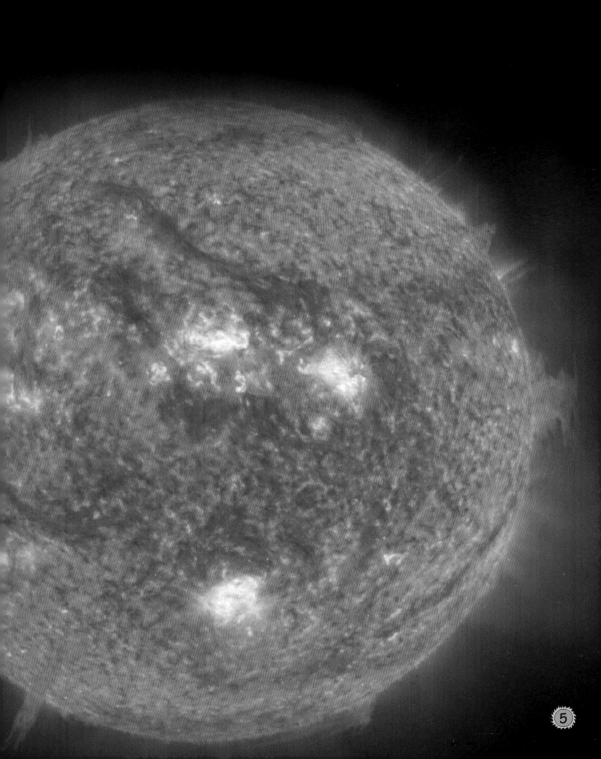

太阳系 The Solar System

太阳位于一个庞大天体家族的中心。

The Sun is at the center of a large family of space objects.

在这里，所有的天体都绕着太阳公转。

All these objects are **orbiting**, or circling, around the Sun.

其中就有八大行星。

There are eight planets orbiting the Sun.

它们分别是水星、金星、我们的母星地球、火星、木星、土星、天王星和海王星。

The planets are called Mercury, Venus, our home planet Earth, Mars, Jupiter, Saturn, Uranus, and Neptune.

冰冻的彗星和岩质小行星也绕着太阳公转。

Icy **comets** and rocky **asteroids** are also circling around the Sun.

太阳、行星和其他天体共同组成了"太阳系"。

Together, the Sun, the planets, and other space objects are called the **solar system**.

小行星是太空中的巨大岩石。大多数小行星都集中在被称为"小行星带"的环状带中，绕太阳旋转。

Asteroids are huge space rocks. Most of the asteroids orbiting the Sun are in a ring called the asteroid belt.

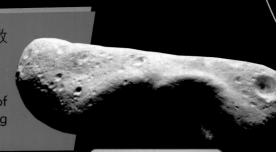

小行星 An asteroid

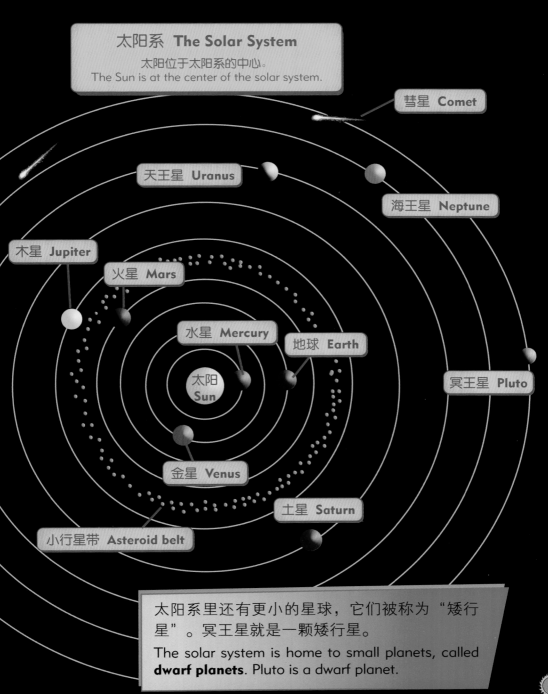

太阳系 **The Solar System**

太阳位于太阳系的中心。
The Sun is at the center of the solar system.

彗星 **Comet**

天王星 **Uranus**

海王星 **Neptune**

木星 **Jupiter**

火星 **Mars**

水星 **Mercury**

地球 **Earth**

太阳
Sun

冥王星 **Pluto**

金星 **Venus**

土星 **Saturn**

小行星带 **Asteroid belt**

太阳系里还有更小的星球，它们被称为"矮行星"。冥王星就是一颗矮行星。

The solar system is home to small planets, called **dwarf planets**. Pluto is a dwarf planet.

太阳是什么?
What Is the Sun?

当你仰望夜空的时候，会看到成百上千颗恒星。

When you look at the night sky, you see hundreds and hundreds of **stars**.

我们的太阳也是一颗恒星，就像其他所有恒星一样！

Our Sun is also a star, just like all those others!

相比于别的恒星，太阳在天空中显得很大，这是因为我们离它比较近。

The Sun looks huge in the sky compared to other stars because we are close to it.

天空中其他恒星与地球的距离，都远远超过太阳与地球的距离。

All the other stars in the sky are much farther from Earth than the Sun.

和所有恒星一样，太阳也是一颗巨大的气体球。

Like all stars, the Sun is a giant ball of **gases**.

随着气体燃烧，它们产生了光和热。

As the gases burn, they create light and heat.

地球离太阳有1.5亿千米。但太阳真的太大了，就算是从地球上看，它在天空中也还是很大。

Earth is 150 million kilometers from the Sun. The Sun is so huge, however, that from Earth it still looks big in the sky.

太阳 **The Sun**

太阳 **The Sun**

冥王星表面
Surface of Pluto

这张图片显示了站在冥王星上看太阳的样子。冥王星到太阳的距离比地球到太阳的距离远得多。从冥王星上看，太阳像是天空中一颗闪亮的耀眼恒星。

This picture shows how it might look to stand on Pluto. Pluto is much farther from the Sun than Earth. From Pluto, the Sun would look like a bright star shining in the sky.

白天与黑夜
Day and Night

在白天的时候，我们能在天空中看到太阳，但是到了夜里它就消失了。

During the day, we see the Sun in the sky, but at night it disappears.

这是因为我们的地球在自转，就像陀螺一样。

This happens because our Earth is **rotating**, or spinning, like a top.

随着地球自转，我们这颗行星的不同地方会经历白天与黑夜。

As Earth rotates, different parts of our planet have day and night.

当你住的地方朝着太阳的时候，你那里就是白天。

When the place where you live faces toward the Sun, it is daytime for you.

当它从太阳的光线中转开以后，黑暗就会降临，变成夜晚。

As it spins away from the Sun's light, darkness falls and it is night.

太阳 The Sun

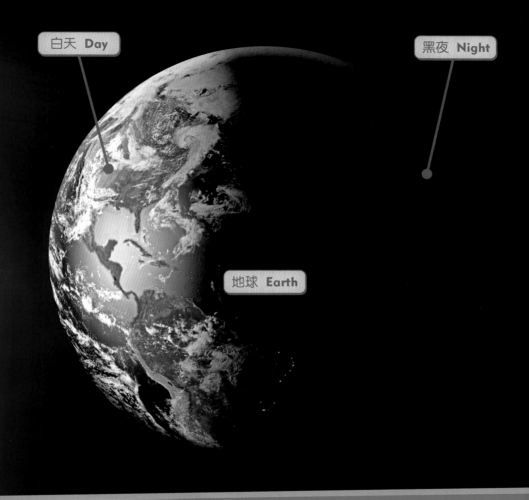

白天 Day

黑夜 Night

地球 Earth

这张图片显示了地球上的白天与黑夜。朝向太阳的半边处于白天，而另外半边则处于黑夜。当然，在现实生活中，太阳和地球可远没有这么近。

This picture shows how day and night look on Earth. It's daytime on the half of the planet that's facing the Sun. On the other half, it's nighttime. In real life, the Sun and Earth are not this close.

近距离观察太阳
A Closer Look at the Sun

太阳和太阳系中所有其他天体都在宇宙中运动。

The Sun and all the other objects in the solar system are moving through space.

当它们在宇宙中运行的时候，太阳也会自转，就像地球一样。

As it travels through space, the Sun is also rotating, just as Earth does.

地球自转一圈要花24个小时，也就是一个地球天。

It takes Earth 24 hours, or one Earth day, to rotate once.

巨大的太阳则需要花上25个地球天来自转一次。

The giant Sun takes about 25 Earth days to spin around once.

与地球相比，太阳是个庞然大物。

Compared to Earth, the Sun is enormous.

可以说太阳里能装下一百万个地球！

It would be possible to fit 1 million Earths inside the Sun!

太阳 Sun

这张图片告诉我们太阳是如何自转的。

This picture shows how the Sun rotates.

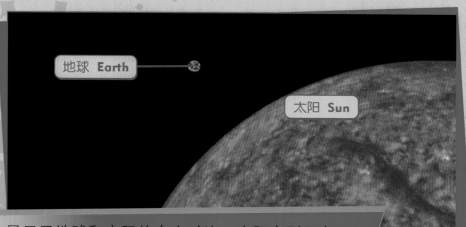

这张图片显示了地球和太阳的大小对比。太阳大到只有一小部分能放在这张图片中。

This picture shows the size of Earth compared to the Sun. Only a small part of the Sun can fit into the picture.

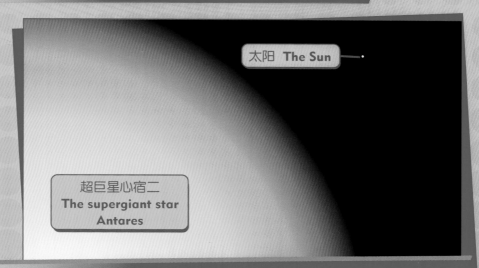

有的恒星比太阳还要大得多。这张图片显示了太阳和超巨星心宿（xiù）二的大小对比。现在太阳看上去只有一点点大了！

Some stars are much bigger than our Sun. This picture shows the size of the Sun compared to a supergiant star called Antares (an-TAHR-eez). Now the Sun looks tiny!

了不起的恒星
Our Amazing Star

当科学家使用特殊仪器观测太阳时，他们看到了很多奇妙的现象。

As scientists watch the Sun using special equipment, they see many amazing things.

有时候好像太阳上面发生过爆炸。

Sometimes it looks as if there has been an explosion on the Sun.

这些突然出现的闪烁被称为太阳耀斑。

These sudden bright flashes are called solar flares.

有时候巨大的超热气体环会从太阳表面喷出。

Sometimes giant loops of super-hot gas erupt from the Sun.

这些灼热的气体环会延伸到宇宙空间中，甚至能长达数十万千米。

These glowing loops can stretch into space for hundreds of thousands of kilometers.

太阳耀斑 Solar flare

这是太阳耀斑的照片。
This is a photo of a solar flare.

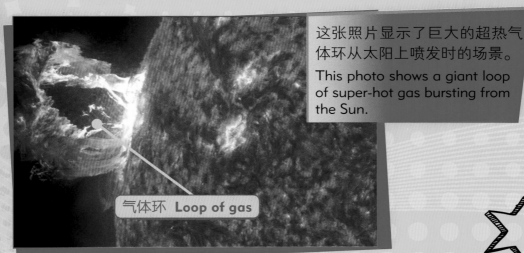

这张照片显示了巨大的超热气体环从太阳上喷发时的场景。

This photo shows a giant loop of super-hot gas bursting from the Sun.

气体环 **Loop of gas**

太阳黑子 **Sunspot**

这张照片显示了太阳表面的黑子。太阳黑子是太阳表面比周围区域温度更低的区域。这张图片上最大的太阳黑子比地球还要大。

This photo shows dark sunspots on the Sun. A sunspot is a place that's cooler than the rest of the Sun's surface. The largest sunspot in this photo is bigger than Earth.

太阳去哪了?
Where Did the Sun Go?

地球绕着太阳公转，月亮则绕着地球公转。

As Earth orbits the Sun, the Moon is orbiting Earth.

有时候，太阳、地球和月亮会恰好处于适当的位置，然后产生奇妙的现象。

Sometimes, the Sun, Earth, and Moon are in just the right position to make something exciting happen.

我们会看到月亮穿过地球和太阳之间。

We see the Moon pass between Earth and the Sun.

当月亮走到太阳面前，它就会挡住太阳的光。

As the Moon passes across the face of the Sun, it blocks the Sun's light.

这叫作日食。

This is called an **eclipse**.

在日食的时候，天空甚至会变得很暗。

The sky may even get darker during an eclipse.

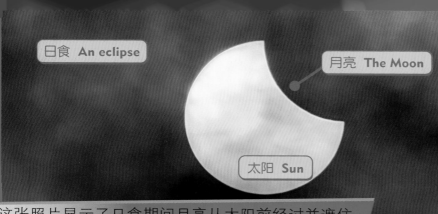

日食 An eclipse

月亮 The Moon

太阳 Sun

这张照片显示了日食期间月亮从太阳前经过并遮住太阳的样子。

This photo shows the Moon passing in front of the Sun during an eclipse.

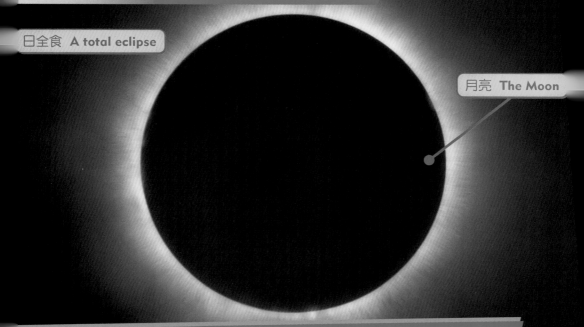

日全食 A total eclipse

月亮 The Moon

有时候月亮会完全遮挡住太阳，这就叫作日全食。在日全食期间，白天会变成黑夜，但是只会持续几分钟。

Sometimes the Moon completely blocks the Sun. This is called a total eclipse. During a total eclipse, day seems to turn into night for just a few minutes.

研究我们的恒星
Studying Our Star

人类难以直接造访炽热的太阳，但是我们可以通过空间探测器来展开研究。

Humans cannot visit the burning-hot Sun, but we can study it using space **probes**.

一个叫作太阳动力学天文台（SDO）的空间探测器正在对太阳进行研究。

A space probe known as SDO is studying the Sun right now.

SDO在距离地球表面大约36 000千米的高空绕地球自转。

SDO is orbiting Earth about 36,000 kilometers above Earth's surface.

它几乎每秒都会向地球传输一张太阳照片。

The probe beams a photo of the Sun back to Earth every second.

如果没有太阳，我们都无法在地球上生存。

We couldn't survive on Earth without our Sun.

所以尽我们所能去研究这颗对我们来说意义非凡的恒星是非常重要的。

So it's important that we learn everything we can about our very special star.

太阳动力学天文台(SDO)
SDO probe

这张照片显示了探测任务开始前科学家们在SDO周围工作时的样子。

This photo shows scientists working on SDO before its mission began.

科学家们 Scientists

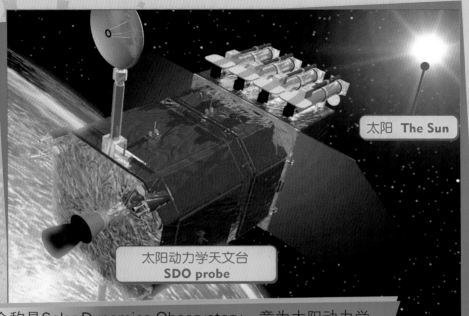

太阳 The Sun

太阳动力学天文台
SDO probe

SDO的全称是Solar Dynamics Observatory，意为太阳动力学天文台。这张图片模拟了观测站在地球上空飞行时的样子。

The name SDO is short for "Solar Dynamics Observatory". This picture shows how SDO might look as it flies above Earth.

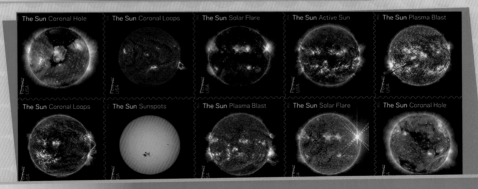

SDO已经拍摄了上亿张太阳图像。2021年，美国邮政局发行了一系列邮票来展示这些美丽的图像。

SDO has captured hundreds of millions of images of the Sun. In 2021, the United States Post Office released a set of stamps showing these images.

有趣的太阳知识
The Sun Fact File

以下是一些有趣的太阳知识。　　Here are some key facts about the Sun.

太阳是如何得名的
How the Sun got its name

古罗马人称太阳为"Sol"，在英语中"Sol"的意思是太阳。
The Romans called the Sun "Sol". In English, the word "Sol" means "Sun".

太阳的大小
The Sun's size

太阳的直径约
1 391 016千米

1,391,016 km
across

太阳自转一圈需要多长时间
How long it takes for the Sun to rotate once

大约25个地球天
About 25 Earth days

太阳的速度
The Sun's speed

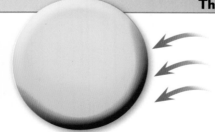

太阳在宇宙中运动的速度大约是790 000千米每小时。

The Sun is traveling through space at about 790,000 km/h.

太阳上的温度
Temperature on the Sun

太阳表面的温度是5 500摄氏度。
The temperature on the Sun's surface is 5,500°C.

太阳与地球之间的距离
The Sun's distance from Earth

地球距太阳的最近距离是147 098 291千米。
地球距太阳的最远距离是152 098 233千米。

The closest Earth gets to the Sun is 147,098,291 km.
The farthest Earth gets from the Sun is 152,098,233 km.

太阳系的大小
Solar system sizes

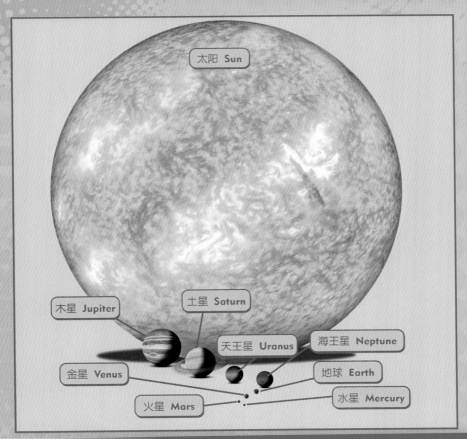

太阳 Sun

木星 Jupiter

土星 Saturn

天王星 Uranus

海王星 Neptune

金星 Venus

地球 Earth

火星 Mars

水星 Mercury

这张图片显示了太阳系中太阳和各行星之间的大小对比。
This picture shows the sizes of the solar system's planets compared to the Sun and each other.

动动手吧：用熔化的蜡笔制作太阳
Get Crafty : Melted Wax Crayon Sun

使用熔化的蜡笔，做一个色彩丰富的"阳光捕手"装饰品吧！

你需要：
- 蜡纸
- 剪刀
- 黄色、红色和橙色的蜡笔
- 奶酪刨
- 熨斗（在成年人的帮助下使用）

1.从蜡纸上剪下两个餐盘大小的圆。

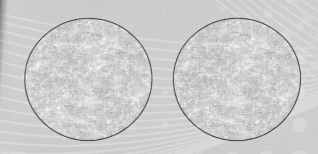

2.使用奶酪刨将蜡笔屑刨在其中一张圆形蜡纸上。（应有成年人在旁监督，务必注意手指安全！）

3.将另一张蜡纸盖在蜡笔屑上，做成蜡笔"三明治"。

4.请成年人帮忙熨烫蜡笔"三明治"，直至蜡笔屑熔化并混合在一起。

5.将熔化的蜡笔太阳图挂在窗前，沐浴在阳光下。或者用它来制作一幅星空图。

词汇表 Glossary

小行星 | asteroid

围绕太阳公转的大块岩石，有些小得辆像汽车，有些大得像座山。

彗星 | comet

由冰、岩石和尘埃组成的天体，围绕太阳公转。

矮行星 | dwarf planet

围绕太阳运行的圆形或近圆形天体，比八大行星小得多。

日食 | eclipse

指太阳的光被月亮挡住。

气体 | gas

无固定形状或大小的物质，如氧气或氨气。

公转 | orbit

围绕另一个天体运行。

行星 | planet

围绕太阳公转的大型天体：一些行星，如地球，主要是由岩石组成的；其他的行星，如木星，主要是由气体和液体组成的。

探测器 | probe

不载人太空飞船。通常被送往行星或其他天体，用于拍摄照片并收集信息，由地球上的科学家操作控制。

自转 | rotate

物体自行旋转的运动。

太阳系 | solar system

太阳和围绕太阳公转的所有天体，如行星及其卫星、小行星和彗星。

恒星 | star

燃烧的巨型气态星球。我们的太阳就是一颗恒星。